THE INSTITUTE AT PATRIOTS POINT

Engineer's Guide to Preventive Maintenance

Mitigating Asset Risks Through Preventive Action

Authored by

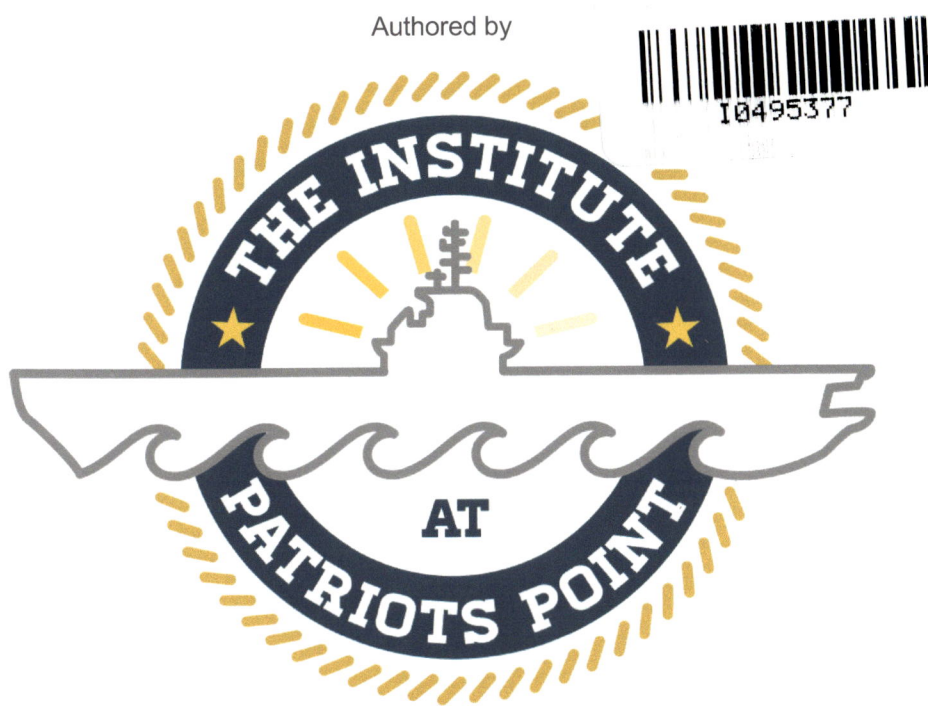

Darrin J. Wikoff
Rick Clonan

Copyright © 2019 Eruditio, LLC
Institute at Patriots Point • 40 Patriots Point Road
Mt. Pleasant, SC 29464
Phone 843.375.8222 • www.theinstituteatpatriotspoint.com

TABLE OF CONTENTS

Introduction 5

Principles of Preventive Action 11
    Defining Preventive Maintenance 11
    Defining Failure 14
    The P to F Interval of Time 17

Preventive Maintenance Development 18
    What Are the Assets? 21
        General Procedure 24
    How Do Assets Fail? 27
        General Procedure 29
    What Decisions Should We Make? 33
        Elements of the Equipment Maintenance Plan 33
        General Procedure 34
        Setting the Task Frequency 36

Preventive Maintenance Optimization 39
        Getting Started 43
        Operator Care 48
        The Importance of Data 50

LIST OF TABLES

Table 1 - Consequence Risks 31
Table 2 - Probability Risks 31
Table 3 - Detectability Risks 31
Table 4 - Equipment Maintenance Plan Asset Elements 34
Table 5 - Equipment Maintenance Plan Task Elements 35
Table 6 - Equipment Maintenance Plan Support Elements 36
Table 7 - PM Task Selection Chart 38
Table 8 - Preventive Maintenance Evaluation Checklist 39
Table 9 - Preventive Maintenance Maturity Assessment 44

PREFACE

ISO 55001 outlines the management system requirements for Asset Management and defines "Preventive Action" as a demonstration of an organization's ability to proactively identify potential non-conformance in asset performance (i.e. failures) and mitigate the risk. This book, as a continuation of Eruditio's series on *Leadership for Asset Management Excellence*, provides Engineers with a guide to document Failure Modes and Effects Analysis as identified asset-related risks and the selected risk mitigation tasks to prevent these non-conformances from impacting asset management objectives.

In collaboration with industry and academic leaders, this book is intended to be used as a resource for evaluating and designing asset management plans for preventive action. The views and perspectives expressed within this resource are those of the authors based on their collective experience as members of the Society for Maintenance and Reliability Professionals (SMRP), the U.S. Technical Advisory Group to ISO PC-251, and as instructors and community leaders in asset management.

About the Authors

Richard "Rick" Clonan is a U.S. Navy submarine veteran with more than 28 years of maintenance experience, specializing in Preventive Maintenance program development and optimization, EAM implementation, and maintenance workflow process improvement. Rick is a Certified Maintenance and Reliability Professional (CMRP) and holds an RMIC Certification from the University of Tennessee. He currently serves as the Best Practices Committee Chair for the Society of Maintenance and Reliability Professionals (SMRP), and is the Implementation and Training Manager at Eruditio, LLC.

Darrin J. Wikoff specializes in Organizational Change Management, Lean Manufacturing, Business Process Re-Engineering and Reliability Engineering. Since 2001, Darrin has continued to train, coach and mentor industrial leaders through the rigorous process of implementing and managing improvement initiatives in support of Lean Manufacturing. He has also authored *"Centered On Excellence"* published by MRO-Zone in 2012 and the 7th edition of the *"Maintenance Engineering Handbook"* published by McGraw-Hill in 2008.

INTRODUCTION

One of the greatest challenges facing modern industry in a post-millennial economy is access to capital. "Capital" refers to the financial resources available to a company. Capital also refers to the financial value of things, such as assets, machinery, real estate or even intellectual property. Studies conducted since 2012 indicate that access to capital is among the top three risks to businesses that are operating in a growth market.

Capital is different from money. Money is used to purchase the goods and services needed to support the business. Capital is used to generate a future financial value (e.g. "wealth") for the business like an investment. In the context of assets, the future financial value of an item is unrealized until the company applies the necessary labor and material resources needed to transform the physical equipment into money.

The business of reliability, asset management or maintenance is to ensure that each asset owned by the company realizes its full potential value and does so without incurring additional risk. As such, and in the context of capital, the business of reliability must translate organizational objectives into technical and financial decisions, plans and day-to-day activities that convert the investment made in physical assets to money or cash.

First, let's define the term "Asset". An asset, with regards to capital, is any item that has a quantifiable value to the business and its stakeholders. An asset, relative to reliability, must also have a quantifiably distinct function or purpose within the business. Using the automobile as our example, value is quantifiable in terms of the purchase price, down payment, resale value, or the loan amount for the vehicle. The automobile has a quantifiable value, in terms of both money and capital, but we might also divide the vehicle into smaller groupings of assets based on their uniquely different functions. An "Asset Hierarchy" within our Enterprise Asset Management systems (EAM) illustrates the relationship of assets within both the financial and reliability realms.

Tires, for example, as a replaceable item will require additional investment from the vehicle's user over the life of the parent asset, the automobile. As such, the tires may also be an "asset", either as a grouped system of components or as individual items. Because the tires can be used over more than one fiscal or financial reporting period, they are a capitalized item, an "asset". The cost of replacing the tires serves as the initial value needed to define the asset.

For our second definition, relative to reliability, if we assume that the function or purpose of the tires, in the context of the parent asset, is to provide friction between the drive train and the road to achieve 85% power transmission efficiency, then the asset formerly known as "tires" should be divided into a minimum of two unique assets. In terms of the power transmission efficiency function, the two rear tires, on a rear-wheel drive automobile, would be classified as a single asset. Because you need both rear tires to create the required level of power transmission efficiency, the asset is "Rear Tires". This grouping of components within the asset definition is known as a "System". The system is the unique asset and a single asset identification number would be assigned to the "Rear Tires". Because the front tires do not enhance or impede the power transmission efficiency function, they are not part of the "Rear Tires" asset. As such, the front tires must have a different purpose or function as a system, or as individual assets, and would be assigned an asset identification number different than the "Rear Tires".

Once we understand what assets exist within our business, and we have defined the quantifiable values and functions of each asset, we shift our attention towards identifying the risks associated with each asset's ability to transform capital into cash. This is, after all, the reason why we made the investment in the first place. Managing the risks associated with physical assets requires a focus in two parallel areas: Effect – the consequence associated with the asset not performing its required function as desired – and Probability – the level of uncertainty that the asset will perform its required function as desired.

The easier of the two focus areas to begin with is the discussion regarding the Effect a non-performing asset has on the business. In the setting of the automobile, the most significant effect would be based on your inability to drive the car to and from your place of employment as the user or primary stakeholder. Quantifying the Effect, in this case, would be determined by the amount of time you are unable to drive the vehicle and the income you lost within this period. Risk, in these terms, is measured as:

$$Effect = \frac{\text{Consequence}}{\text{Time}}$$

However, since you are not the only stakeholder associated with the value of the automobile, we must also consider other risks. If you secured a loan to purchase the vehicle, then the Bank may be a stakeholder. The Bank's risk is based on you not being able to make your loan payment on time while you are unable to use the vehicle to generate an income. The interest lost on the Bank's asset would replace your income in the numerator of the risk equation.

$$Effect = \frac{\text{Lost Loan Interest}}{\text{Time}}$$

In general, when quantifying the Effect risk associated with an asset you must consider the following:

- What is the impact to the primary stakeholder or user?
- What is the impact to other financial stakeholders?
- What is the impact to the physical environment in which the asset operates?
- What is the impact to other assets in the value stream?

The "Probability" risk associated with an asset refers to the level of uncertainty that an asset will perform its function as desired. Probability is a measure of confidence in your ability to ensure value – convert the capital investment into cash. Uncertainty, because of not knowing what will go wrong or what may cause the asset to fail to meet stakeholder expectations, is the risk we are trying to quantify with metrics like Overall

Equipment Effectiveness (OEE), Availability (Ai), Probability of Reliability (Rt), and Mean-Time-Between-Failures (MTBF) as a leading indicator.

$$Probability = \frac{Uncertainty}{Time}$$

Probability, in comparison to the Effect of asset non-conformance, is variable based on the operating context associated with how the asset is being used. Returning to our automobile analogy to help us understand these core principles, would the level of uncertainty increase if you changed how you operated the vehicle? Assume your normal operating mode is to drive 20 miles from home to work and back again five days every week. How certain are you that the vehicle will get you from point 'A' to point 'B' without breaking down on the side of the road? Pretty confident I'm sure. You make sure each day that you have enough fuel to make the trip, and occasionally have your vehicle serviced to ensure it is in good working order.

Now, assume that you are taking a 1,400-mile road trip from Dallas, Texas to Los Angeles, California across the deserts of Arizona and through the steep mountain passes in eastern California. Would you be as confident in your preparations and current asset management plans? Would you decide to take extra preventive actions to ensure your 1,400-mile trip was successful? Sure, you would! You might have the tires rotated, balanced and inflated to a lesser pressure to compensate for the extreme climates. You might even purchase a new spare tire and extra fuel tank knowing that the 200-mile journey through the desert is without frequent service stations. What about you as the Operator of the automobile. Would you change your driving practices too? You might set a cruising speed of 65 mph to optimize fuel economy or make frequent stops to check your tires and other mechanical components that are subject to fatigue in the extreme climates. All these technical and financial decisions are based on the increasing level of uncertainty resulting from the new operating mode. The function of the asset hasn't change. You want to travel from point 'A' to point 'B' without breaking down. The consequences haven't changed. Any breakdown of the automobile will result in the same Effect, loss of income, increase in expenses, or loss of usability in general. The only risk parameter that has changed based on the operating context is the level of uncertainty, or probability of a breakdown over the 1,400-mile journey (e.g. "time").

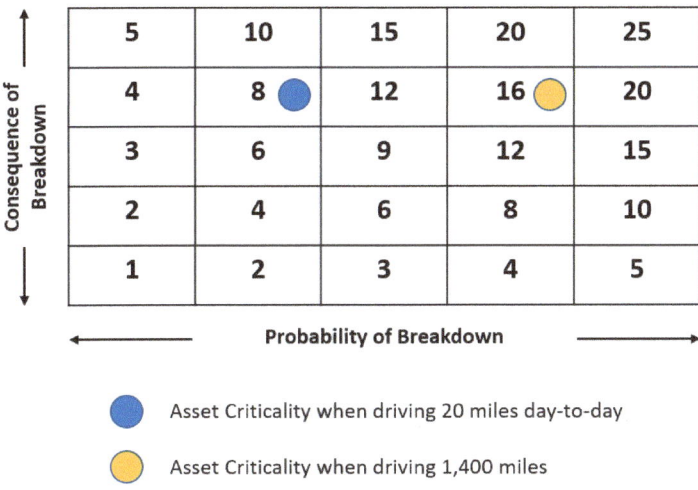

Figure 1 - Operating Context Asset Criticality Matrix

If assets exist within a business to increase the business' ability to generate revenue, then our asset management plans and activities should center on mitigating the risk to said value. The last asset management principle we are going to discuss in this introduction deals with the approach to asset management planning. Asset management plans define how a business ensures value. The genesis of each plan to operate or maintain is derived from the identified and quantified level of risk deemed unacceptable by stakeholders. Asset management plans must clearly link the identified business risks to prescribed activities or tasks and the resources required to accomplish each activity. These plans are not limited to countermeasures as reactionary devices used to recover once an undesirable effect has occurred. Instead, these plans are focused on the prevention of risk all together, by reducing the consequences of non-conformance or the probability of non-conformance, or both.

Asset management planning requires a systematic approach to ensure that all feasible causes and effects are identified, and that the business has provisioned for those activities that will mitigate the risk to a suitable level. Asset management planning typically follows four steps:

1. Risk Analysis – Evaluating each operating mode to determine the causes of undesirable effects. This is most commonly referred to as "Asset Criticality Analysis".
2. Task Selection – Identifying the operating or maintenance practice that will prevent the undesirable effect, or in the least predict the conditions that will result in the undesirable effect if unmanaged. "Failure Modes and Effects Analysis" is a frequently used practice.
3. Resource Planning – Determining the skills, competencies and informational requirements necessary to consistently execute the prescribed tasks. An "Equipment Maintenance Plan" provides this level of detail.
4. Budgeting – Allocating the financial resources required by the resource plans. Maintenance labor and material budgets should be defined annually to ensure proper resource allocation.

PRINCIPLES OF PREVENTIVE ACTION

A Preventive Maintenance program (PM) is a list of tasks that are to be performed on a portfolio of assets for the purposes of identifying defects that could result in asset failures or the potential loss of function (i.e. non-conformances). These activities are typically preformed at a scheduled, routine frequency, and can include minor adjustments, replacement of components, visual or quantitative inspections, testing of components or asset performance, calibration of components that serve as a point of product judgement, or lubrication of wearable mechanical components.

The frequency of PM tasks can be time-based, calendar-based, or interval-based using an hour meter or similar cycle counting measurement system. Each of these frequencies does not take into consideration the condition of the asset or component being inspected and thus relies heavily on failure and performance data analysis to be effective.

- Time-based inspections - the task is to be performed at a fixed frequency based on the characteristic life of a wearable component.
- Interval-based or "meter-based" tasks - are performed at a predetermined point in the asset's lifecycle based on usage, like the number of cycles of a press.
- Calendar-based tasks - are those tasks that are performed based on the time of year, like an annual overhaul or testing freeze protection during the winter months.

Defining Preventive Maintenance

Preventive Maintenance activities cover a very wide spectrum of maintenance actions, however, there are some maintenance activities that often get incorrectly included as "preventive maintenance". So, what determines what is and is not a PM activity? Very simply, it comes down to timing. When is the maintenance task performed within the asset lifecycle? If the activity happens after a loss of asset function, then that activity is not a preventive action.

> *"Preventive Action" refers to those maintenance activities that identify or mitigate a failure, or the effects of a failure, prior to a loss of asset function.*

A maintenance intervention that occurs after functional failure may use the same tools, techniques or procedures as a preventive maintenance task but is not considered a preventive action. "Preventive Action" refers to those maintenance activities that identify or mitigate a failure, or the effects of a failure, prior to a loss of asset function.

There are several reasons why Preventive Maintenance (PM) is a popular strategy in today's modern industrial settings. The purpose of PM is to maintain or restore asset function as defined by the needs of the business. Specifically, a good PM program will identify potential failures of components before the loss of asset function results in downtime, process inefficiency or product quality losses. PM activities also allow Maintenance organizations more time to plan their work, source their materials, and coordinate corrective action without adding risk to the business.

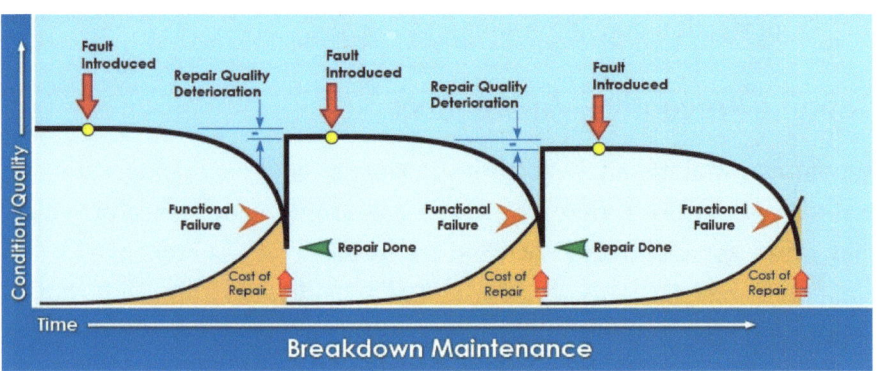

Figure 2 - Breakdown Maintenance Cycle

The diagram in Figure 2 illustrates the impact of breakdown maintenance using John Moubray's "P to F" management philosophy. In his book, *RCM II*, John Moubray explains that a complex asset has multiple points, over time, when the warning signs of failure could have prevented a loss of functional performance if detected. Notice the high cost of repair for breakdown maintenance in Figure 2, when the fault was introduced, and the potential point of detection that could have alerted the organization in time to complete a corrective action before loss of function. Also note that a reduction in repair quality will lower the overall condition of the asset when put back into service. This will result in an overall reduction in the operation of the asset because the condition or quality of the asset function will not be as high as it was before the fault occurred. This is why it is so important to have well written procedures that prevent the introduction of defects during maintenance and define the acceptance testing criteria needed to verify that the installed asset capability (e.g. "function") has been restored. In the absence of these procedures, it is possible that a repair can be made that does not restore the asset back to functional specifications.

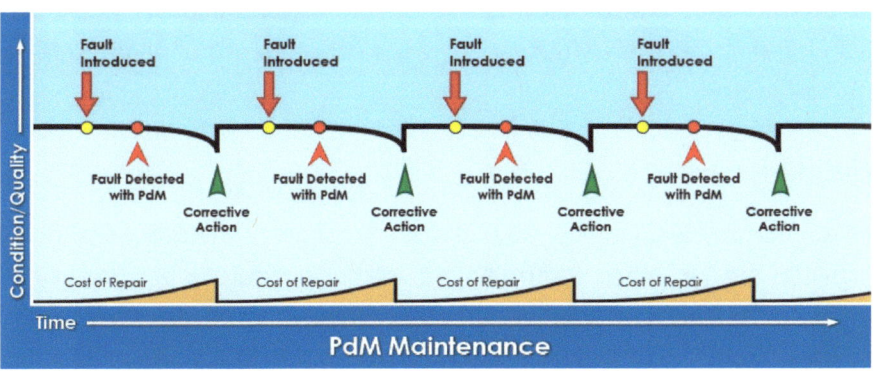

Figure 3 - Predictive Maintenance (PdM) Cycle

Figure 3 shows what happens when proper PM inspections take place and the impact of preventive action over time. The fault is introduced in the same location, but due to predictive failure mode detection techniques, a corrective maintenance action can be taken to reduce the downtime impact on Production Operations. These activities are often less expensive than waiting until the asset ceases to function as required due to the reduced defect severity – fewer damaged or degraded components.

"Predictive Maintenance" is a type of preventive maintenance that is designed to evaluate changes in the health or condition of an asset based on specific failure modes. Although scheduled using the same PM frequencies, predictive maintenance (PdM) routines are meant to proactively identify random failure mode effects. In the predictive maintenance cycle (Figure 3) the number of maintenance interventions is higher within the same period of time, but, due to the cost of repair decreasing, the total maintenance cost per asset will be less than the breakdown cycle by at least 3:1. Defects resulting from failure mode effects are detected earlier in the P to F interval of time, thus isolating the defect to a single component or part.

Some Preventive Maintenance tasks are designed to replace components prior to the point where they are most likely to fail. Others, like vibration analysis or infrared thermography, are predictive in nature and allow asset conditions to be monitored over time to more effectively detect the warning signs of failure, thus reducing the consequence of failure side of the risk matrix.

Defining Failure

To this point, we have explained that Preventive Maintenance (PM) activities are in place to determine the presence of defects or prevent the effects of failure, but what is meant by "failure" with regards to asset management?

Failure occurs when a component no longer does what we want it to do. Typically, machines don't fail, rather a component stops performing its intended function and the effect is seen as a reduction or loss of the machine's function. Since assets are made up of many components, each component having several failure modes – a specific manner of failure – the opportunity for functional failure at the machine-assembly level is quite high if nothing is done to prevent it. Some components wear out over time, but most just simply give up and stop working because of a change in operation, a change in the operating environment, or due to an underlying defect that was introduced during installation or maintenance. This is known as a "random", non-age-related failure pattern. Finding the right point in time when the random change in asset condition is most evident and applying the proper preventive maintenance action to detect it, is preventive maintenance Nirvana.

Prior to 1978, as part of a Federal Aviation Administration (FAA) task force, Stanley Nowlan, Howard Heap and other executives and engineers from commercial aviation, the United States Airforce and other regulatory commissions studied failures in the aviation industry and they discovered that aircraft components fail predominantly in six distinct patterns as shown in Figure 4.

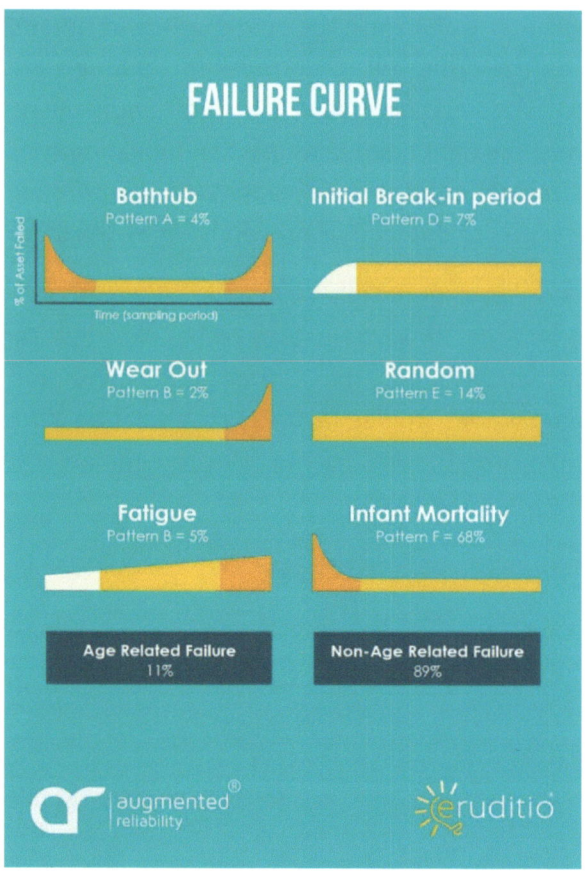

Figure 4 – Six Predominant Failure Curves

Out of the thousands of components examined during the study, and the numerous operating modes of a Boeing 747, it was determined that the failure patterns on the left showed an indication of wear out, or an increasing probability of failure over time based on the number of assets, historically, that had failed within given time intervals. In the FAA study, the Task Force examined component failures at 2,000 flight hours, 5,000 and 10,000 flight hours. "Wear Out" is indicated in these curves by an increasing percentage of assets in a failed state on the right side of each failure pattern. It was also their conclusion that time-based preventive maintenance, such as time-based inspections, time-based replacements, or time-based overhauls, are best suited for age-related failure patterns like these.

However, there are a clear majority of failure modes that are not effectively managed by time-based tasks as the only means of preventive action. As illustrated in the failure patterns on the right, and specifically the flat horizontal line that indicates random failure probability, traditional preventive maintenance intervention will not typically have a measurable impact on non-age related failures, and tasks directed at preventing random failures through time-based interventions should be scrutinized heavily for effectiveness.

The P to F Interval of Time

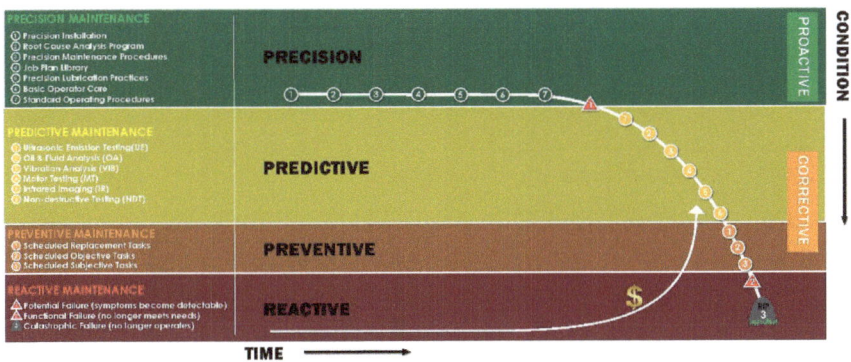

Figure 5 - P to F Curve and Maintenance Domains

The "P to F Curve", as an interval of time, is a visual representation of the rate of degradation of a machine assembly based on the numerous failure modes associated with the components that make up the machine. It is often incorrectly associated with the "failure pattern", which is assigned at the component level. Remember, assets are made up of many components. And, components have multiple failure modes that are predictable using the component's predominant failure pattern. One component can degrade differently over time than that exact same component used in a different operating context. But, both like components will exhibit common failure modes as they degrade, and thus a common failure pattern. For instance, a bearing in a dry clean area may not require lubrication as often as that exact same bearing used on a washer conveyor where it is exposed to warm water for most of its life. Both will show the warning signs resulting from insufficient lubrication, but they may do so at different intervals of time.

What the "P to F Curve" represents is the duration of time between the detection of a potential defect, 'P', and failure, 'F'. It illustrates that there are many options for us to choose from to detect the presence of seemingly random failure modes prior to recognizing the effects of machine failure, and the proper point of preventive action that effectively mitigates the machine's loss of function. For example, the duration of time between detection of a bearing defect with vibration analysis and functional failure of the pump is greater that the duration of time between detection of a drive belt defect using a time-based visual inspection method and pump failure. Using this same analogy, if the bearing defect is undetected, the drive shaft becomes misaligned due to the eccentric rotation caused by the bearing defect, and the drive belt will begin to wear along the sheave at an accelerated rate. The bearing defect exhibits a random failure pattern while the belt is still a wear out pattern. As a machine, the pump assembly begins to show signs of functional failure as the severity of defect transitions from the bearing, to the shaft and eventually the drive belt. We will return to the P to F Curve again later when we discuss choosing the proper intervention method.

PREVENTIVE MAINTENANCE DEVELOPMENT

Understanding the physics of failure is most important when creating a Preventive Maintenance (PM) program from the ground up. Components display specific failure modes, and these failure modes can be modeled as a predominant pattern of failure for the component. Within each machine assembly, like a pump, there are several components. The effects of failure within one component will induce a defect or failure mode in another component, until the machine assembly itself degrades to the point of functional failure.

Now we need to turn our attention towards PM Development (PMD). Figure 6 outlines the application of the same Reliability Centered Maintenance (RCM) methodologies documented by Stanley Nowlan,

Howard Heap and John Moubray as the pioneers of RCM. Each methodology serves an explicit purpose during the development of a PM program. PM Development should not solely focus on creating new preventive action job plans. It must also create a pathway to routinely evaluate and optimize PM routines.

We will organize our PM development discussion in three key areas, and provide a detailed procedure to accomplish each objective:

1. Asset Hierarchy Development – systematically grouping assets (e.g. components, machine assemblies and subsystems) relative to the common function or purpose they serve within the plant or business.

2. Failure Mode Analysis and Coding – methodically evaluating the common and specific manner of failure that defines each component's failure pattern and establishing a data structure to monitor and trend the frequency of each pattern.

3. Equipment Maintenance Planning – documenting preventive actions, resource requirements and budgetary considerations for the purpose of communicating these plans to all stakeholders.

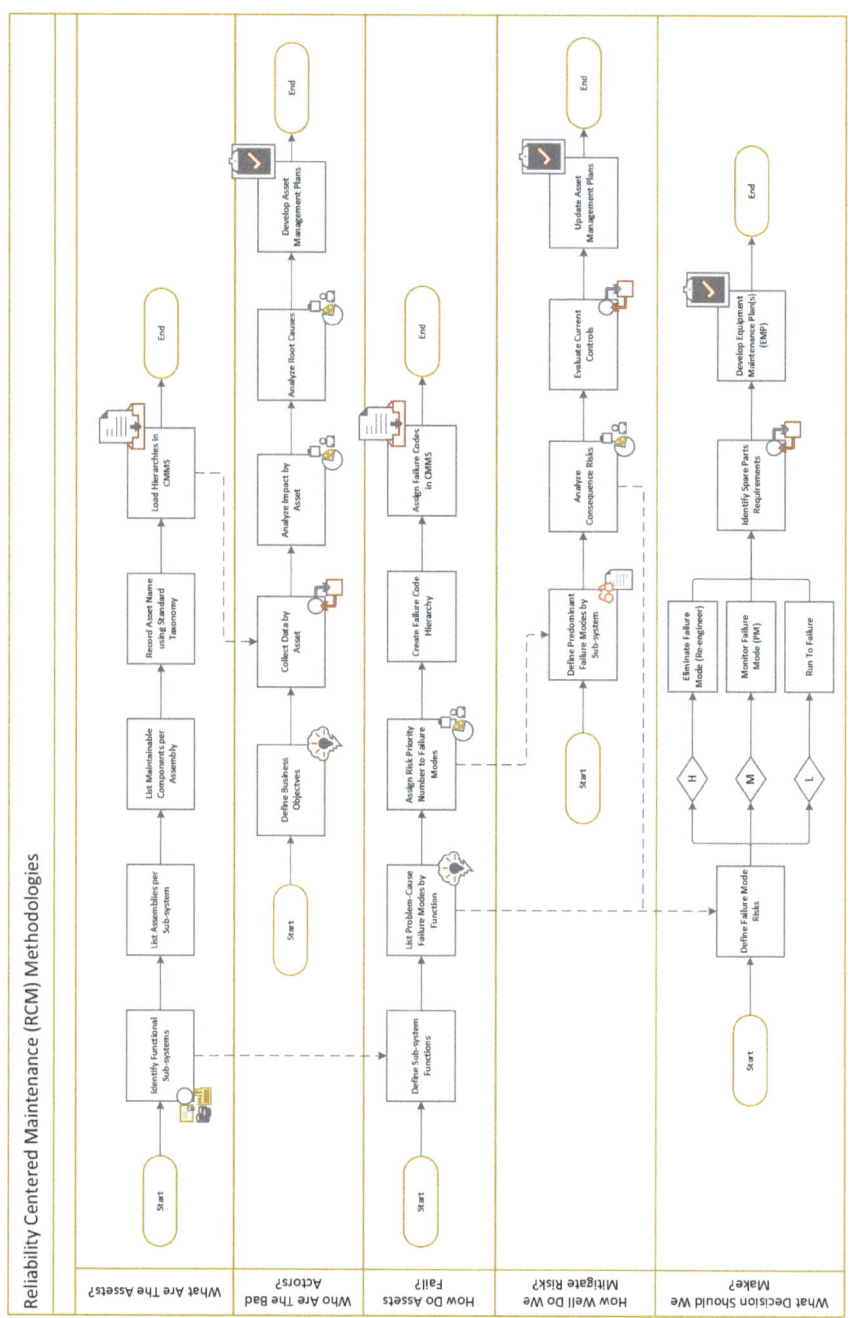

Figure 6 - Reliability Centered Maintenance Methods

What Are the Assets?

"Hierarchy" is defined as a series of ordered groupings of things within a system. In the case of reliability engineering, this refers to the physical equipment, machinery, and related structures within a manufacturing or production facility that are used to process or manufacture finished products or support that function.

The **Asset Hierarchy** is a series of ordered groupings of *Assets* in a *Parent-Child Relationship*.

Reference Standard	Standard of Practice	Objective
ISO 14224:2016 - Petroleum, Petrochemical and Natural Gas Industries - Collection and Exchange of Reliability and Maintenance Data for Equipment	A parent-child functional hierarchy exists in the CMMS, beginning at the critical system level and cascading down to the lowest maintainable component within each system.	To improve maintenance work order cost distribution, and more readily display asset failure history at the lowest maintainable level.

Asset is defined as any item that has a distinct and quantifiable *Function*. Assets can be physical equipment, a building, a capitalized spare, or any other item that is uniquely different from other items in your plant or facility.

Function is a manner of operation that adds value to the business. The asset's function is what we want the asset to do, like milling a part at a rate of 17 per hour, within tolerance, to sustain a production takt time of 2.7 days.

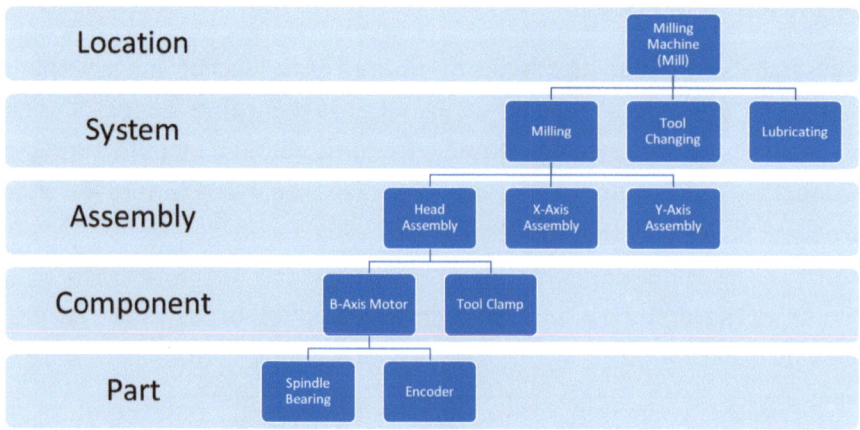

Figure 7 - Asset Hierarchical Structure

The **Parent-Child Relationship** refers to the bond between levels of the asset hierarchy. In the Figure 7 example for the Milling Machine, the Head Assembly is a child of the Milling system, which is a child of the Location, the Mill itself. If the Head Assembly fails, so does the milling system and, subsequently, the Mill fails to meet the desired function of 17 per hour. However, a failure of the Head Assembly does not impact the Mills' ability to position the tool in the 'X' or 'Y' axis. We know this because the "X-Axis" and "Y-Axis" assemblies are also children of the milling system, at the same indenture level as the "Head Assembly". The impact or effects of failure are equally consequential, but independent of each other.

The **Location** of an asset in your Computerized Maintenance Management System (CMMS) is used to facilitate Work Order tracking, Labor Reporting, Preventive Maintenance administration, Inventory administration, Item Masters, Asset Hierarchies, Service Requests, and other applications. The Location code, like "Milling Machine", in your asset hierarchy also defines the link between physical asset hierarchy and general ledger accounts in your financial accounting systems. Before we can create physical assets, we first need to determine the Location in the CMMS where labor, inventory and other costs will be reported against the general ledger. The cost of maintaining the "Head Assembly" rolls up to the cost of "Milling" and is recorded at the "Milling Machine" location in the general ledger.

A **System** is a grouping of assets that serve a common function. For example, in a Mill we group one combination of assets together that are used for milling, and group other combinations of assets together that are used for tool changing, lubricating, or controlling the Mill. We separate these asset groups, or systems, so we can collect data specific to each function of the Mill.

An asset **Assembly** (e.g. "Subsystem" or "Equipment Unit") is a collection of two or more assets that are grouped together based on how they support or serve the system's function. Relative to the Mill, the "Head Assembly" includes the "B-Axis Motor" and 'Tool Clamp" components, as well as the "Spindle Bearing" and "Encoder" parts. Alone, the Head Assembly cannot mill parts at the desired rate. However, we want to capture failures and the cost of maintaining the Head Assembly as a subset of the milling system. This makes it easier to see how the Head Assembly is performing in relation to other assets associated with the Mill.

The **Component** within an asset assembly is the lowest level of the asset hierarchy at which we want to administer work orders and collect work history for analysis. Within the "Head Assembly" of the Mill, the "B-Axis Motor" would be a maintainable component. We would then analyze the specific modes of B-Axis Motor failure and develop preventive maintenance job plans to identify and eventually correct B-Axis Motor defects. Some software systems prohibit component-level asset hierarchies. Using "Failure Codes", discussed in the subsequent section *How Do Assets Fail*, failure history can be reported and trended at the component level even if a component asset record doesn't exist. However, component-level assets should be created to facilitate work order assignment, cost distribution and other work histories at a lower level of data granularity.

A **Part** is like a component, except we do not plan to write a work order to a part, and we will not collect work history for the part. We will, nevertheless, track how often the part is issued or used by the Storeroom using a Stock Keeping Unit (e.g. Stock Number), instead of an Asset ID number. Within the "B-Axis Motor" of the "Head Assembly", the "Spindle Bearing" would be a part. There's no right or wrong answer. The decision is solely based on whether you want to collect work order and failure history at the component level. In many organizations the "Spindle

Bearing" is a component, while others choose to simply record the frequency at which bearings are purchased and issued from the Storeroom.

General Procedure

Step 1: Asset Selection
Asset Selection is a process by which the Engineer works with the customer in developing the criteria needed to determine which assets will be included in the PM Development scope of work.

Begin by setting the focus area boundaries. Obtain a copy of the current Value Stream Map, or similar document, that defines the manufacturing or utility process flow. Identify the system-level groupings of assets included in the asset scope of work.

Then, develop the criteria that will be used to select critical systems. This is the criteria that will be used to select the assets for further evaluation within the project. It is important to understand that the goal of criticality analysis is to produce a granular ranking of assets. There needs to be differentiation in the results. If the results do not clearly identify the top ten percent of critical assets, without question, there will always be a lack of trust in the analysis and poor utilization of asset criticality results within maintenance workflow processes. The criteria used to rank asset criticality should be granular enough to clearly sort the important from unimportant systems. Examples of criteria include:

- Health, Safety and Environmental impact,
- Asset value threshold,
- Maintenance cost threshold,
- Production impact threshold, and
- Regulatory requirements

Next, review and select the critical systems. Only those assets classified as critical systems will move to the next step in the process. A "critical system" is the reliability engineering term used to define the level of probability and consequence risks, as described in our introduction. Ranking system-level assets relative to the highest inherent risk to the business is the preferred method of finalizing PM Development priorities.

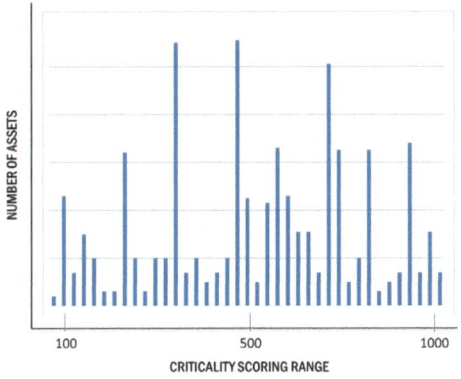

Figure 8 - Criticality Analysis Histogram

A Histogram chart can be used to illustrate the distribution of criticality analysis scores for each system. Scores should be randomly distributed across the entire scoring range. A modal or tailed distribution indicates that the analysis criteria is not suitable for all assets, or the analysis criteria lacks sufficient granularity to differentiate the important from the unimportant.

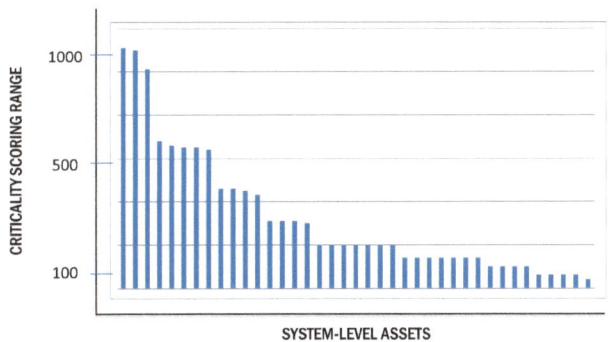

Figure 9 - Criticality Analysis Pareto Chart

A Pareto chart is used to illustrate the priority of system-level assets by sorting assets from the highest individual score to the lowest. Those assets that fall within the top 10% of the scoring range should be selected for PM Development.

Step 2: Define Asset Classifications

All assets should be assigned to a unique "Asset Class" based on their function. Defining asset classifications increases efficiency when creating failure codes, preventive maintenance job plans and performing failure analysis. Examples of asset classifications are:

- Air Handling Unit
- Cooling Tower
- Precision Milling Machine
- Centrifugal Pump

Step 3: Asset Hierarchy Creation

Export the existing hierarchy contained within your CMMS. This information should contain attribute data such as asset name, location, equipment type, manufacturer and model if available. Create the initial hierarchy from the exported data as your starting point.

Within your straw model hierarchy, define the property accounting hierarchy. Obtain a copy of the accounting general ledger that defines capital assets or "property" owned by the business. Populate the "Type of Installation", "Plant Location" and "Location" levels of the asset hierarchy using the property nomenclature and general ledger account codes (e.g. Asset ID Tag or General Ledger ID).

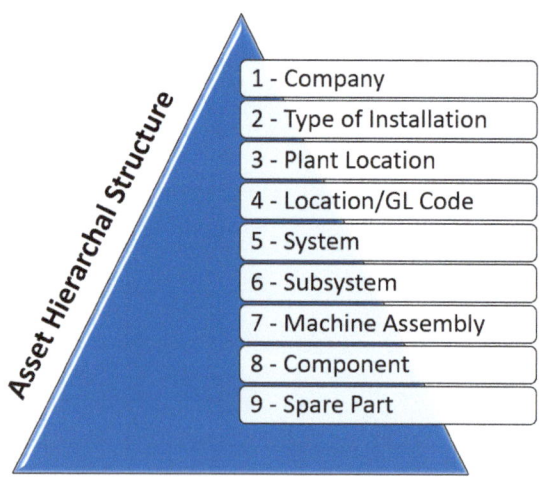

Figure 10 - Asset Hierarchal Structure

Now, create the functional hierarchy Location ID. The Location level of the hierarchy in your CMMS defines the link between physical asset hierarchy and general ledger accounts in your financial accounting systems. This is the upper-most level of the functional hierarchy. Examples include:

- Air Handling Unit #1
- Cooling Tower #1
- FOG Mill #1
- Fresh Water Pump #1

Using the exported data, list the parent-child relationships for each level of the asset hierarchy. Refer to Figures 7 and 10 as examples. Assign an "Asset Class" to the system, subsystem and machine assembly levels of the asset hierarchy.

Finally, verify the Asset Hierarchy by walking down the system to capture undocumented assets, verify asset relationships, and capture attribute date required for failure mode analysis (i.e. motor rpm, number of greased bearings, drive belt size, HMI ID number, etc.).

How Do Assets Fail?

The **Failure Code Hierarchy** application in your Computerized Maintenance Management System (CMMS) is used to build a failure hierarchy to record, track and trend the failure of assets, by asset classifications. The failure hierarchy shows the relationships between identified problems, causes and remedies or actions performed by maintenance for equipment-based and location-based failures. Once the failure hierarchies have been built, maintenance can record, report and analyze equipment failures more easily as a method of routine PM evaluation and optimization, without the need to manually review each Corrective Maintenance (CM) work order.

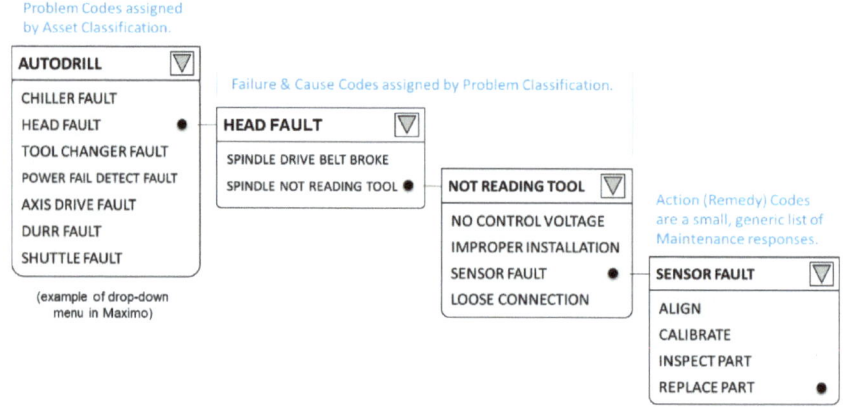

Figure 11 - Failure Code Hierarchy Example

Problem Codes identify the general problem associated with an asset, such as "leaking," "high temperature," or "won't start." This coding structure, organized by asset class, enables more detailed reporting and allows Maintenance Technicians to easily find the code they need.

Failure Codes define the specific component that failed and the type of failure, "shaft bent," "bearing worn," or "gasket leaking," for example. These codes introduce more granularity into failure reporting, allowing the Maintenance department to analyze equipment failures with greater detail.

Cause Codes identify the general reason why the asset was underperforming, such as "design flaw," "operator error," or "misalignment." These codes are designed to be used by a Reliability Engineer as a starting point for Root Cause Analysis (RCA).

Action Codes, sometimes called "remedy codes," identify the action that maintenance took to rectify the equipment issue. Because these codes describe general maintenance activities, such as "cleaned" or "replaced," they are generically defined and apply to all asset classes and failure codes.

Reference Standard	Standard of Practice	Objective
ISO 14224:2016 - Petroleum, Petrochemical and Natural Gas Industries - Collection and Exchange of Reliability and Maintenance Data for Equipment	Equipment failures are recorded on work orders by choosing the appropriate failure codes from the failure hierarchy drilldown menus.	Improve granularity of failure reporting, allowing Maintenance and Engineering to analyze equipment failures using automated reporting tools.

General Procedure

Step 1: Failure Modes and Effects Analysis

The goal of this procedure is to identify the <u>common</u> failure modes associated with critical systems to enable a failure mode-based maintenance strategy and failure code hierarchy.

As a starting point, a failure mode review is used to map asset classifications for the selected critical systems to known failure modes. This data may be provided using a component failure mode library or previously completed Failure Modes and Effects Analysis (FMEA). The objective in this first step is to identify those asset classes with incomplete data, or where the level of uncertainty is greatest.

Next, review your work order history. Export from your CMMS the past 36 months of work order history. Sort the list of completed work orders by asset classification. Generate a Pareto Chart for the sorted list by failure frequency, corrective action frequency, or even the frequency of spare parts consumption. This step helps us prioritize asset classes needing a more detailed failure mode analysis and ensures that our efforts are <u>focused on the things that have happened</u>, and not the things that might happen. By using Pareto Analysis to identify the most common failure codes for each asset class, based on maintenance history, this accomplishes a few things:

- It establishes a ranking of the most probable failures by asset class,
- It provides a justification for changing the current preventive maintenance routines, and
- It gives us a baseline to review the effectiveness of our PM Development efforts.

Failure Mode Analysis is where we complete the explanation of, "How do assets fail?". This process should focus on the most frequently occurring asset classification failures as the initial priority. Within a cross-functional focus group, perform the following:

a. Define Functions - Write or list the function of each assembly within the asset hierarchy. A function statement should include rate, speed, pressure, distance or some form of pass-fail measurement.

b. Create a Functional Block Diagram - Draw a functional block diagram that illustrates the relationships of assembly functions within the parent system. List the components associated with each functional block (e.g. machine assembly).

c. List Failure Modes - In terms of the "Component", "Failure" and "Cause", list the reasons why each assembly fails to meet its desired function. Examples are:
    - Axis / Overtravel Alarm / Sensor Fault
    - Motor / High Temperature Alarm / No Coolant
    - Bearing / High Temperature Alarm / No Lubricant
    - Pump / Low Flow / Fluid Restriction

d. Sort by Risk Priority Number (RPN) - Assign a risk priority number to each failure mode that quantifies the Severity of functional failure, the Probability of failure mode occurrence, and Detectability of the failure mode prior to system-level functional failure. Risk priority numbers should clearly delineate the important from the unimportant. Use Tables 1-3 to help you get started. Sort failure modes from highest to lowest risk.

Table 1 - Consequence Risks

Effect	Severity of Effect	Rank
Catastrophic	The failure mode effects safe operation and may cause death or injury or creates a non-compliance with government regulation without warning.	5
High	Item inoperable, with loss of primary function. No safety, environmental or regulatory impact.	4
Medium	Item operable, but at a reduced level of performance.	3
Low	Item operable, with a workaround. Customer impact is minimized. No loss of system-level function.	2
None	No effect to Operations/Customer. Qualified personnel can prevent failure consequences.	1*

Table 2 - Probability Risks

Failure Rates	Probability of Failure	Rank
> 1 in 1 month	Very High: Failure is inevitable	10
1 in 2 months		9
1 in 4 months	High: Repeatable failure	8
1 in 6 months		7
1 in 1 year	Moderate: Occasional failure	6
1 in 2 years		5
1 in 3 years		4
1 in 5 years	Low: Relatively infrequent failure	3
1 in 8 years		2
< 1 in 10 years	Remote: Failure is unlikely	1*

Table 3 - Detectability Risks

Control	Detectability of Cause	Rank
Undetectable	Current controls will not or cannot detect the system-level functional failure.	5
Unlikely	Current controls will detect functional failure.	4
Moderate	A Preventive Maintenance inspection will detect the potential cause <u>before</u> functional failure occurs.	3
Likely	An Alarm will detect the potential cause <u>before</u> functional failure occurs.	2
Certain	Automated controls (e.g. PLC System) will prevent the potential cause of failure.	1*

Step 2: Failure Code Development

For meaningful failure mode analysis, and to easily evaluate PM effectiveness, the number of unique "Failure Codes" per asset classification should be limited. Failure code development should focus on the <u>most common</u>, or highest risk priority failure modes first. Routine audits of completed work orders will help refine failure code hierarchies over time. The use of "Other" or "Miscellaneous" codes is not recommended unless the failure reporting fields in your CMMS are mandatory fields for work order closeout. In the event of mandatory fields, a metric should be established that reconciles the percent of work orders completed with "Other" failure codes. A threshold of 10% or greater should trigger an audit by Engineering to refine failure code hierarchies.

Create a Failure Code Hierarchy using the failure modes collected during the failure modes and effects analysis. Capture the following for each asset classification:

- Problem Codes - Capture the functional failures reported by the FMEA team. "Problems" should reflect the end user's point of view and terminology. For example, "Won't Rotate" or "Won't Start".
- Failure Codes - Record the "Failure" within each failure mode documented by the FMEA team. "Failure" should be specific to an asset classification, such as "Axis Overtravel" or "Bearing High Temp" and includes the specific component or part in the description.
- Cause Codes - Record the "Cause" of each failure mode documented by the FMEA team. "Cause" codes should echo a specific manner of failure, like "No Coolant" or "Fluid Restriction", to add granularity to future failure Root Cause Analysis.

What Decisions Should We Make?

The Equipment Maintenance Plan, or EMP, is the document Maintenance and Reliability Engineers use to communicate the asset management plan to Maintenance Planners and Schedulers who are responsible to develop standard maintenance procedures and schedule the preventive maintenance work orders within the maintenance workflow process.

An EMP can be created for a specific asset, as is commonly done when performing a failure modes and effects analysis, a general equipment type or asset classification, such as centrifugal pumps or AC motors greater than 25 horsepower, or to communicate the engineered strategy for an entire system, like the Milling Machine (Mill). When creating your EMP it's important to first decide how the EMP will be used and if your strategy will be applicable to other areas or assets within the overall facility or enterprise. If you have several assets that have the same function, and therefore functional failures and failure modes, then the equipment type EMP would be appropriate. This will allow you to create one engineered strategy and deploy it multiple times. If your area of engineering responsibility is focused on a single production line or critical system, and this system has a limited number of assets that are unique to your area only, then it's advisable to build an EMP for the system.

Elements of the Equipment Maintenance Plan

The Equipment Maintenance Plan not only identifies the specific tasks necessary to minimize failure mode consequences, it also contains information that guides work order scheduling and provides pertinent information for budgeting maintenance resources and communicating downtime requirements to Production Operations. The elements of the EMP can be broken down into three categories of information: Asset Elements, Task Elements and Support Elements. We will demonstrate how to record this information in four steps as shown in Figure 12.

Reference Standard	Standard of Practice	Objective
ISO 55001:2014 – Management System for Asset Management - provides principles, framework and a process for managing asset-related risks.	Asset Management Plans are formally documented for each function within the asset management system, including design, storage, installation, operation and maintenance.	To demonstrate that maintenance plans are derived from a risk-based methodology and specifies the type of preventive action required and the frequency at which the task will be performed.

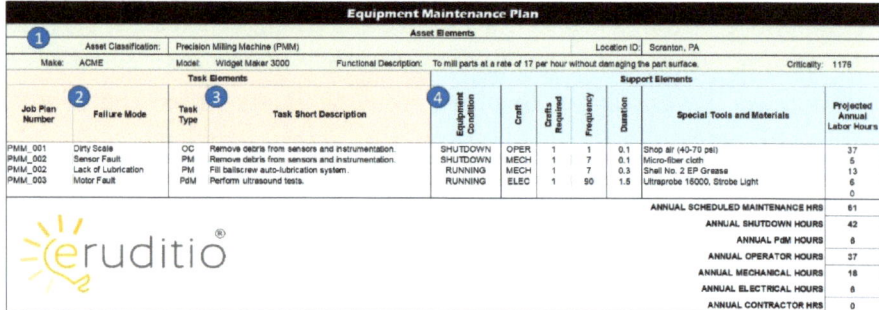

Figure 12 - Equipment Maintenance Plan

General Procedure

Step 1: Record the "Asset Elements", including a function statement and asset classification (e.g. "Equipment Type").

Table 4 - Equipment Maintenance Plan Asset Elements

Equipment Name or Type	The name or type of equipment to which the EMP applies. This could include several different models and manufactures, but the functionality of the equipment is the same. For example, your equipment type might be a "Rotary-Screw Chiller".
Equipment Criticality Ranking	The ranking or rating given to the equipment based on the results of your criticality analysis. If you are building the EMP for the entire system, then you would use the system level criticality ranking.
Manufacturer and Model	Include all the equipment models to which the EMP applies. This will help Planners and Schedulers identify the assets within the computerize maintenance management system.
Function and Functional Failure	List the performance expectations, or functions of the component, and the ways in which the component fails to meet the required function.
Component	The nomenclature used within your computerized maintenance management system to identify the specific asset for which the maintenance task applies. This will further improve the Planner and Scheduler's ability to create the required work orders.

Step 2: Copy failure modes from the *Failure Modes and Effects Analysis*. Recording of failure modes within the EMP satisfies regulatory requirements that demand documentation of risks and risk mitigating actions, such as ISO 9001, ISO 14001 and ISO 55001, among others.

Step 3: Record the "Task Elements" as the documentation of preventive action.

Table 5 - Equipment Maintenance Plan Task Elements

Maintenance Requirement Number	A unique alpha/numeric number that identifies each task within the EMP. This will be used by Planners and those resources responsible for the development of standard maintenance procedures. The Maintenance Requirements Number will serve as link between the procedure and the EMP.
Failure Mode	The reasons why the component fails to meet expectations, and the basis for prescribing the maintenance task within the EMP.
Task Type	Describes the actual maintenance activity that will be performed on the equipment. Typically, the types of tasks prescribed within an EMP are limited to the following: • PM – Quantifiable preventive maintenance inspections, lubrications or calibrations. • PdM – Predictive routines such as vibration data collection and analysis, infrared thermography, oil sampling and analysis and ultrasonic emissions testing. • CBM or SIT – Condition-based or situational requirements that are performed when a specific defect or failure mode is identified. These types of tasks are commonly referred to as PM or PdM follow-up corrective actions. This will allow you to link the inspection to the corrective action to evaluate the effectiveness of your engineered strategy.
Task Description	The purpose of the task. For example, "Collect vibration data for analysis", or "Inspect belt-drive for visible defects". The task description is not the procedure; it only defines the expected outcome of the task.

Step 4: Record the "Support Elements" and estimated labor-hours required to complete each task as defined in Table 6. The support elements serve as documentation of the resource requirements needed to administer and sustain your asset management plan.

Table 6 - Equipment Maintenance Plan Support Elements

Equipment Condition	This indicates if the equipment needs to be shutdown or if it should be running in a normal mode of operation when performing maintenance requirement task.
Craft	Ideally this element would identify the skill level of the individual performing the task, such as 'Mechanical Level II'. However, at a minimum you should record the craft type, like 'MECH' for mechanical or 'ELEC' for electrical. It is also appropriate at this point in the EMP to identify those tasks that are best performed by an Operator using the code 'OPER', or contractor using 'CNTR'.
Crafts Required	This is the estimated number of maintenance resources required to complete the task. Planners will further clarify the resource requirements when scheduling the work order.
Frequency	The periodicity of the maintenance requirement, such as weekly, monthly, quarterly, etc. If your computerized maintenance management system uses days, hours or meter-hours, to set the frequency of which work orders are generated it's advisable at this stage to follow the same structure to prevent confusion during scheduling. It is recommended that you also determine if the frequency of the task will be "floating" – the next work order will be scheduled once the existing work order has been completed in the work order system – or "fixed" – the next work order will be scheduled once the defined period of time (e.g. 30 days) as elapsed, regardless of when the last work order was completed.
Duration	The estimated labor-hours needed to perform the maintenance requirements. This should not include work preparation time, lockout/tagout time, or other indirect time associated with the work execution process. These durations will be added to the work order by the Planner or Scheduler as they may vary by area or the specific operating context of each asset for which the EMP applies.
Materials and Tools	In an effort to help Planner's and others build the equipment Bill of Materials, and to ensure standardization of spare parts and tools used to execute the engineered strategy, you should include the materials and special tools associated with each task on the EMP. "Special Tools" are those tools that are not frequently used by craft persons and may include items like torque wrenches, calipers, micrometers and rigging that require additional training before use, or may be limited in numbers and controlled by the storeroom or "Tool Crib".

Setting the Task Frequency

To ensure that Preventive Maintenance routines are adding value, corrective action is required. Inspections detect the presence of failure modes, but eventually you must correct the known defects to improve reliability and maintainability. When setting the task frequency for randomly distributed failure patterns, like bearings and electronic devices, the preferred frequency is 1 Standard Deviation from the Mean-Time-Between-Failures (MTBF). This ensures a PM to CM ratio of 6:1.

The "6:1 Rule" is more than a rule of thumb. It is a statistical evaluation of preventive maintenance effectiveness and provides an early warning indication of too little detail in the PM job plan, training deficiencies, the wrong PM frequency, or the wrong method being used to detect the failure mode characteristics associated with the asset.

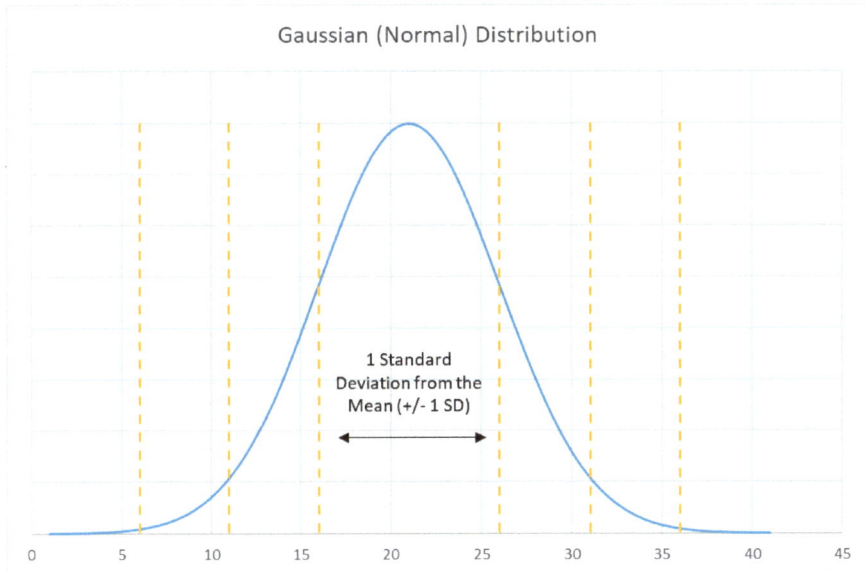

Figure 13 - Normal Distribution Curve (e.g. "Bell Curve")

Figure 13 demonstrates the normal distribution of random failures, with three standard deviations to the left and right of the peak of the curve, the "Mean" – central tendency or average of the data. Within the three standard deviations exists 99.73% of all the reported failures over the analyzed period. If we assume that the Preventive Maintenance (PM) work orders are scheduled at a frequency equal to one standard deviation from the mean, then we should find evidence of a failure mode once for every six times the PM is completed. "6" in the 6:1 Rule refers to the six PM inspection opportunities across the normal distribution of failures, and the "1" states that one Corrective Maintenance (CM) work order was generated within the same time interval.

To determine the proper task, we must establish two parameters: characteristic life (Eta) and failure rate (Beta). The "Beta Value" identifies the rate at which each component fails over its characteristic life, or "Eta Value". Eta is the equivalent point in time at which 63.2% of like items will fail.

A beta equal to 1 (β=1) represents a predominantly random failure pattern. For these components, a continuous condition monitoring task should be selected for early defect identification. By continuous, instead of routine monitoring, we mean that the inspection frequency is more often than 1 Standard Deviation from the asset's MTBF. For example, an asset with a beta of 8 would be inspected routinely every six months, approximately, if it's MTBF is 8,760 hours or 1 year. However, an asset with a beta of 1 would be equipped with on-line condition monitoring or a predictive technology would be deployed monthly to trend dynamic changes in the asset's condition. Monthly vibration monitoring of bearings is a common strategy since ball, roller and sleeved journal bearings typically have a beta of 1. A beta of 1.1 to 1.9 is still "1". Although bearings should last for 150,000 to 450,000 service hours, monthly monitoring is required due to the predominantly random introduction of defects caused by misalignment, over or under lubrication, improper loading and contamination.

Table 7 - PM Task Selection Chart

Failure Pattern	Beta Value	Select technology to monitor changes in asset conditions.	Select time-based inspection method to quantify wear rate.	Prescribe time-based replacement plan at 75% of characteristic life.	Prescribe post-installation acceptance test to eliminate defects.
Random	$\beta = 1$	✓			
Infant Mortality	$\beta < 1$	✓			✓
Slow Aging (Fatigue)	$1 < \beta < 6$		✓		
Wear Out	$\beta > 12$			✓	

A beta value that is less than 1 (β<1) demonstrates an infant mortality failure pattern. This type of failure pattern requires a continuous condition monitoring strategy because most failures will occur randomly after an initial break-in period. Rigorous post-installation inspections are recommended to ensure defects aren't introduced prior to operation. Standard Operating Practices, or SOP's, should also be reviewed to ensure that frequent starting and stopping of the asset doesn't contribute to an increase in failure rate and decrease in characteristic life.

By now we should understand that a beta greater than 1 ($\beta>1$) is indicative of a wear out failure pattern, but our maintenance plans may vary depending on the size of the Beta value. A beta greater than 12 ($\beta>12$) is a high confidence wear out failure pattern and is best managed using a time-based replacement strategy. A beta greater than 1 but less than 6 ($1<\beta<6$) requires a quantitative asset health inspection to validate that the asset has entered the "wear out" period. The inspection method may vary from intrusive, quantitative inspections using gauges or meters to non-intrusive predictive technologies like vibration analysis or oil analysis based on the severity of risk.

PREVENTIVE MAINTENANCE OPTIMIZATION

PM Optimization (PMO) is the process by which routine PM tasks are evaluated to determine their effectiveness. The goal of PMO is to remove non-value adding tasks or replace intrusive inspection methods with condition monitoring or predictive technologies (e.g. "Predictive Maintenance") to improve maintenance labor utilization and reduce asset downtime. Table 8 provides a checklist for evaluating PM tasks.

Table 8 - Preventive Maintenance Evaluation Checklist

Failure Mode Based

	The preventive maintenance task detects a failure mode identified by your Failure Modes and Effects Analysis (FMEA).
	The preventive maintenance task monitors a known physical parameter or condition (i.e. displacement, acoustic emissions, or changes in pressure, temperature, flow).
	Work order history confirms the need for preventive maintenance. The failure mode is evident in corrective or emergent work orders over the past 36 months.

Scheduled at the Correct Interval

	The preventive maintenance interval is 1 standard deviation from the Mean Time Between Failures (MTBF).
	The preventive maintenance task generates 1 Corrective Maintenance (CM) work order for every 6 times the PM work order is completed.

Organized for Efficiency

	The preventive maintenance task list is organized into a logical sequence that facilitates efficient execution. Did you walk the job plan?
	The preventive maintenance task description uses a verb-noun action statement (i.e. Lubricate the drive end bearing...).
	The preventive maintenance task includes the task duration and Craft Code (i.e. Lube Technician, Mechanic III, General Electrician, HVAC Technician).

Repeatable

	The preventive maintenance task uses quantifiable measures to evaluate asset health (i.e. gauge, caliper, meter) without subjectivity.
	The preventive maintenance task is written without ambiguity. Various maintenance technicians interpret the task objective the same way.
	The preventive maintenance task includes criteria to determine healthy or unhealthy asset conditions.

PMO can be triggered when the asset criticality changes, when Engineering initiates a change to the design of the asset, or the PM strategy for the asset is perceived to be ineffective because unscheduled maintenance continues to occur, even though PM work order compliance is acceptable. The latter is usually identified through work order and failure history analysis.

Asset criticality can change if the operational context of the asset changes, as we explained in our introduction, or a physical configuration change occurs through redesign or new capital installations. Depending on the criticality analysis factors, something as simple as spare parts availability can change the ranking. Asset criticality should be reviewed at least once every two years, and if the review causes an asset to move into the top 10% to 20% of the scoring range, then PMO is highly recommended.

If there are maintenance activities happening on a critical asset on a scheduled basis, but the asset continues to have issues meeting the production requirements due to poor uptime, quality, or safety, especially if the failure modes are detectable, then PMO is a necessity. This is one of the most common triggers for PMO activities and can often have surprising results. Tasks that were believed to be value-added may be removed and new failure detection opportunities may be identified. For example, some tasks could be replaced by condition monitoring techniques to save time and money while simultaneously improving detection results and eliminating unnecessary asset downtime.

The first step in PMO is to gather the necessary data to start the analysis. Those necessary items include:

- The asset criticality analysis results,
- Original Equipment Manufacturer (OEM) manuals,
- An export of the current PM job plans from your Computerized Maintenance Management System (CMMS),
- Corrective maintenance work order history for the past 36 months,
- Lubrication or operator care routines and their frequencies,
- The asset's Bill of Materials (BOM),
- Mechanical and electrical drawings, if available, or a recent equipment walkdown record, and
- The Failure Mode Analysis documents, including the most recent FMEA.

Every PM task should be evaluated when performing PMO, including those that are regulatory in nature. PMO can be completed by asking a series of questions. The first one is "Does this task attempt to address a failure mode that will cause the asset to lose function?" If the answer is no, then the task can probably be eliminated unless it is regulatory driven, administrative or required to comply with plant safety or environmental controls. With FMEA data, it is very simple to see what failure modes should be addressed. If there is an attempt in the task to address a consequential failure mode, then keep the task and proceed through with your PM evaluation using the checklist provided in Table 8.

Cost is also a factor that should be considered when evaluating PM tasks. Is the cost of the inspection justified by the estimated or previously demonstrated breakdown cost? For example, performing a $1,000 inspection every thirty days to prevent a $500 breakdown twice a year is probably not cost effective. The task can be analyzed to lower the inspection cost or eliminated and replaced it with a situational corrective action – a corrective maintenance task that is preplanned and scheduled based on known conditions. Other costs to consider are loss of quality, cost of injuring personnel, and the cost of environmental incidents.

Task frequency analysis takes into consideration asset failure history when available. The consequence of the failure and the estimated rate of failure after detection also play a role in determining task frequency. Task frequency needs to be set to a point that gives the maintenance workflow process enough time to identify, plan, schedule, and execute the corrective action prior to the loss of system function. As stated previously, for random failure patterns, the rule of thumb is to schedule the inspection or monitoring routine at an interval equal to 1 Standard Deviation from the mean-time-between-failures, MTBF, to achieve the 6:1 ratio of PM work orders to CM work orders. For age-related failure patterns, a time-based replacement should be scheduled at 75% of characteristic life.

That brings us to time-based replacements. Does the asset failure data show a wear out pattern that can be mitigated by a scheduled replacement of a component? This form of PM optimization is best used for wear surfaces, fasteners, drive belts, and insulation. Although preventive in nature, time-based replacements are considered proactive "corrective maintenance" and should not be coded as a "PM" work order in your CMMS. The time-based replacements and other similar tasks, although not designed to be failure-finding, are mitigating the loss of function and are, by definition, routine maintenance tasks. Routine tasks are not solely PM work orders. Time-based CM work orders that are managed based on the characteristic life of the component, are considered routine maintenance because replacement is determined without needing an inspection or test to confirm failure.

Finally, let's discuss the regulatory tasks in a little more detail. Typically, when a task is regulated by local or federal authorities there is still some ambiguity on how that task should be carried out. The regulations most often state that a known "non-conformance" (e.g. failure mode) should be mitigated, through preventive action, to a condition deemed acceptable. In the absence of a documented preventive action, regulatory bodies will commonly default to the OEM recommended actions. However, if analysis, like the FMEA, proves the connection between your PM task and the known failure mode, and you have documentation of both preventive and corrective action within your EMP to reduce the identified risks, then the regulatory requirement can and should be adjusted. Check with your regulatory agents before making such a change to your PM tasks.

Getting Started

Optimizing your Preventive Maintenance program may seem like a time-consuming and daunting task. Do not feel as if the only option is to scrap everything and start over. Table 9 maybe a good starting point. The "Preventive Maintenance Maturity Assessment" provides a descriptive method of evaluating the effectiveness of your current PM program for a particular asset or asset class. Start small. Target a few critical systems that are causing the most emotional pain. Review each element of the maturity assessment as a quick justification for further analysis. If you find that your current program falls within "Level 4 – Good Practice" maturity, leave it alone and select another asset.

When assessing program maturity, we need to keep in mind the fact that we are trying to improve both maintainability, as measured using maintenance labor utilization, and asset availability. Program maturity is first evaluated based on maintenance workflow enablement. If "PM Compliance" – the percent of scheduled PM work orders completed on time – is consistently below 60% week-to-week or month-to-month then improvements to PM routines won't have an immediate effect. Your starting point instead should be to improve the maintenance workflow process to increase the volume of PM work order completion. A low PM Compliance may be the result of poor planning and scheduling practices, inadequate spare parts inventories, or more general work execution inefficiencies. Refer to the *Maintenance Manager's Guide to Work Management* for more information about workflow improvements.

Table 9 - Preventive Maintenance Maturity Assessment

	ELEMENTS	LEVEL 3	LEVEL 4	LEVEL 5
Maintenance Workflow Enablement	PM Compliance	Average 60-70%	Average 70-80%	Average 80-90%
	Work Order Creation	PM:CM Ration is > 6:1.	PM:CM Ration is between 3:1 and 6:1.	70%-90% of CM work orders are generated from PM routines.
	Resource Allocation	Dedicated resources to PM program; rarely pulled off task to perform unscheduled maintenance.	30% to 35% Labor Allocation to PM program.	15% PdM and 15% PM Labor Allocation per week.
Preventive Maintenance Program	Failure Mode Driven	Some PM tasks are mapped to the prevention or detection of a specific failure mode.	PM tasks are mapped to the prevention or detection of a specific failure mode that has occurred within the past 36 months.	Level 4, plus < 10% of PM tasks are regulatory or administrative.
	Scheduling Discipline	PM completion occurs within +/- 20% of PM frequency.	PM completion occurs within +/- 10% of PM frequency.	PM completion occurs within +/- 5% of PM frequency.
	Feedback Mechanism	Some PM procedures have craft's feedback when returned; critical feedback is processed by Planner.	50% to 70% of closed PM work orders include Failure Codes.	> 70% of closed PM work orders include Failure Codes.
	Repeatability	Most PM procedures have detailed tasks, steps and some have standard work instructions where needed.	Actual Labor Hours per PM work order are within +/- 15% of Estimated Labor Hours.	Actual Labor Hours per PM task are within +/- 5% of Estimated Labor Hours.
	Quantitative Inspections	Most PM tasks are quantitative with detailed maximum and minimum values.	Level 3, plus PM job plan provides specific instructions for out-of-tolerance readings.	Level 4, plus Engineering evaluates trends for PM optimization.
PM Effectiveness	% Unscheduled Maintenance	< 30%	< 20%	< 10%
	Mean-Time-To-Implement Corrective Actions	> 30 Days	20 to 30 Days	< 20 Days after defect identification.

If the maintenance workflow process is adequately supporting the PM program, our attention then shifts to the integrity of PM routines. We've already discussed these elements, but to summarize, the maturity assessment is designed to highlight those areas of PM Development that may need further analysis. A good PM can be ineffective if it does not consistently provide meaningful data for trend analysis or corrective action selection, or the task and task frequency prevent corrective action in a timely manner – before system functional failure. The "Mean-Time-

To-Implement" (MTTI) corrective action is the leading indicator, impacting the lagging "% Unscheduled Maintenance" indicator.

Let's return to *The P to F Interval of Time* discussion to explain the importance of measuring MTTI. As a measure of time, "P to F" refers to the interval between defect identification (point "P") and functional or catastrophic failure (point "F"). Preventing functional failure – the manner of operation that fails to meet the desired level of asset performance – is the primary objective of preventive action. If this duration of time is 3 weeks (21 days), for example, but it consistently takes the Maintenance workflow process 4 weeks (28 days) to plan, schedule and complete the corrective action that will effectively restore asset health, the PM task and frequency are ineffective in preventing functional failure.

Looking back at Figure 5 – *P to F Curve and Maintenance Domains*, to improve your MTTI, and drive more proactive, corrective maintenance that is planned and scheduled, you must look for opportunities to improve defect identification earlier in the P to F Curve. Using a familiar analogy, a pump assembly, we can explain this philosophy more clearly. For this example, we'll assume that your PM task is "Collect vibration data for analysis" and this task is schedule as a monthly, continuous monitoring frequency based on a failure rate of $\beta = 1$, the predominantly random failure pattern represented by the components that make up the pump assembly. We will also assume that the current MTTI is greater than 30 days and we want to achieve an MTTI of less than 20 days.

In Figure 5, "Vibration analysis (VIB)" is third in the order of predictive maintenance methods, and defects are detectable before functional failure. However, because our MTTI is not within the desired range, our opportunity for optimization is to select "Ultrasonic emission testing (UE)", second in the order of predictive maintenance methods capable of detecting random failure modes associated with the pump assembly. Ultrasonic emission testing is capable of finding bearing and lubrication defects that occur in the predictive domain, but this technology finds the effects of these defects sooner than vibration analysis. Ultrasonic emission testing provides more time to plan, schedule and execute the corrective action, thus extending the P to F interval of time to compensate for our MTTI that is greater than 30 days.

The flowchart illustrated in Figure 14 will help you select the appropriate predictive maintenance technology based on failure mode effects. The flowchart is organized in order of detection priority, with earlier effects (e.g. defects) selected first. If a technology is not available or feasible, quantitative inspection methods should be selected to measure and trend asset health. Quantitative inspection methods include:

- Linear Measurement (caliper, micrometer, dial indicator),
- Gauges (pressure, temperature, flow, multimeter),
- Optical Measurement (microscope, borescope), and
- Image Recorders (motion amplification, high-speed digital camera).

Oil and fluid (lubrication) analysis (OA) detects changes in fluid chemistry, such as viscosity, acidity and the presence of water. This technology may also be used in the "Precision Domain" to detect contaminants that will lead to bearing, gear and hydraulic valve defects if unmitigated.

Ultrasonic emission testing (UE) incorporates a process known as heterodyning to translate ultrasonic frequencies to an audible sound. Failure modes causing an increase in friction or turbulent flow are detectable using this technology.

Vibration analysis (VIB) is most often used to detect anomalies associated with rotating assets but may also apply to vibrating screens or tables and other assets where movement is an undesired effect of failure. This technology captures changes in amplitude – distance, acceleration or force – as the effect of misalignment, imbalance or mechanical looseness, among others.

Infrared thermography (IR) assists anomaly detection by creating a digital thermal image. Failure mode effects that cause a change in surface temperature may be detectable before other, more visible signs of component failure are present.

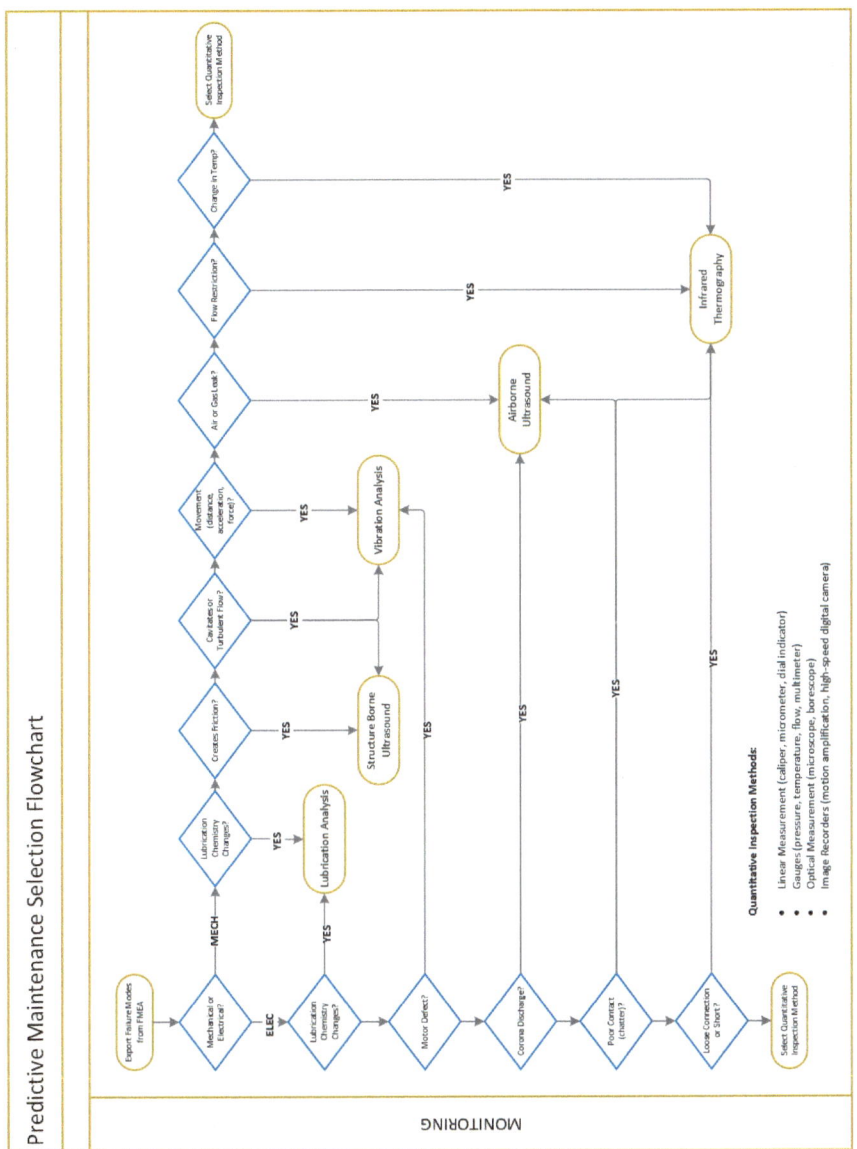

Figure 14 - Predictive Maintenance Selection Flowchart

Operator Care

If a preventive maintenance task doesn't require special tools, special training or technical maintenance knowledge, and can be performed safely by an Operator within their normal job duties, then Operator Care may further optimize the design of the PM program.

Preventive actions performed by Operators increases the available maintenance labor hours per week for corrective action. If we agree that asset health is improved through corrective maintenance, repairing, replacing or refurbishing worn or defective components, then the time maintenance spends performing corrective maintenance is most valuable. Preventive maintenance tasks that can be and should be performed by Operators helps to further optimize the maintenance labor consumed by preventive maintenance.

Operator Care is a simple process that engages all the personnel working within the organization towards a common goal of increased throughput and decreased equipment delays. It is simple in that it engages all our senses in the early identification of equipment abnormalities (rather than the subsequent failures) and provides a simple means to report and track the repairs to be performed. The Operator Care process consists of just a few simple elements designed to ensure that timely and accurate inspections are performed.

Simple, clear, and concise, a Routine Inspection Form defines the Operator Rounds for the area and provides a standard of practice for determining:

- What it to be inspected?
- How often?
- Who is responsible?
- What is the acceptable performance range?
- Where is it located and what does it Look Like?
- And, what should be done if an unacceptable condition exists.

Most importantly, a Routine Inspection Form is developed and maintained by the team working in the area; ensuring ownership and accountability for continuous improvement.

Operator Care is not suitable for complicated maintenance procedures that require special training, skills or tools to complete. Safety of Operators must be considered when determining what should be performed by an Operator. Some of the tasks that could be performed by Operators include:

- Basic Care Cleaning,
- Basic Visual Inspections,
- Minor Adjustments,
- Visual Controls,
- Overall Equipment Effectiveness Reporting,
- Startup and Shutdown Standard Operating Procedures, and
- Asset Validation Procedures.

During PM Optimization, if the above tasks are identified, it is a good practice to see if they fit into the Operator Care guidelines within your plant or facility. Once those guidelines have been established, it is much easier to identify what Operators *could* do. From there, Operations can agree with maintenance what they *will* do. The remaining tasks, if any, will have to be performed by maintenance personnel. Examples of Operator Care restrictions include:

- Use of special tools,
- Requiring special skills or training,
- Use of ladders or lift equipment,
- Task durations extending past shift schedules,
- Task frequencies greater than weekly routines,
- Use of PPE beyond the normal scope of work,
- Use of spare parts beyond the normal scope of work, and
- Tasks requiring coordination with another department.

Once there is agreement between operations and maintenance on what will be done, procedures should be created by a cross-functional team involving Operators and Maintenance Technicians from the area.

The Importance of Data

As stated previously, a Preventive Maintenance Optimization (PMO) activity is often started because the current program is not effective at minimizing downtime or it is believed that the amount of resources applied to the program does not align with organizational goal and is therefore inefficient. In order to prove this, metrics must be in place and utilized properly. This requires accurate and timely data. This data is often captured in the Computerized Maintenance Management System (CMMS) but may be collected from other Enterprise Asset Management systems within a facility, including downtime tracking and Human Resources Management software. Remember, reliability of assets touches every department within an organization.

Some of the data elements that could be leveraged are easily entered on maintenance work orders. Other data may be more difficult to gain access to, such as the cost of downtime per minute or hour, profit margin per unit sold, or even the total available maintenance labor hours per payroll cycle.

It is important to understand that with any change effort, including PMO, it is necessary to show the progress, identify wins, and reward those that are adopting the change as the new way of operating. For this reason, tracking metrics should begin immediately. Start collecting data as soon as possible. The data will probably not be very flattering initially and that is ok. The intent is not to show how poorly an organization is operating. The goal is to show that the program is in fact working and will help those involved in making the change.

In Figure 15, notice how the chart for Downtime Hours indicates the unstable amount of downtime experienced in this facility prior to the project start date, June 1. Within this example, downtime declined after June 1st and stabilized over the rest of the year. This chart shows the results of the resources applied to the new Equipment Maintenance Plans, and the effectiveness of the new PM and other routine maintenance tasks prescribed by the plans.

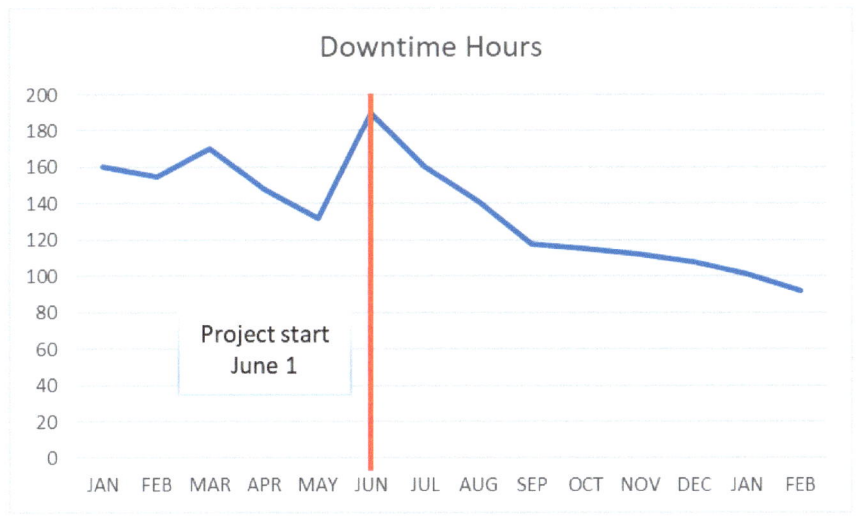

Figure 15 - Data Collection Example

WHAT TO TRACK

Now that the importance of tracking progress is understood, the metrics need to be selected. This can be a very difficult task. It all comes down to the very simple question, "What do we want to improve?" If the issue is downtime is at an unacceptable level, then asset downtime is a good start. If the amount of resources applied is too high, then perhaps a Maintenance Cost per Unit is a good choice. These metrics are high level and cannot be changed directly. It is not possible to simply improve downtime or cost per unit by itself. These metrics are called lagging metrics because they lag other efforts.

Taking downtime as the metric in the example, it is important to determine what activities can reduce downtime. Obviously, having more impactful Preventive Maintenance tasks should reduce downtime. So perhaps tracking Preventive Maintenance Completion rate, Number of Defects Detected, MTTI, and Planned Work Percentage would more immediately indicate the effectiveness of Preventive Maintenance. These types of metrics are called leading metrics because they are an early indication of the program effectiveness and, if done correctly, will result in reduced downtime. It is important to note that metrics do not answer questions but instead get people to ask the correct questions and point focus on the areas that need to be improved. No one, singular metric operates within a vacuum. There are always contributing factors that drive each metrics, and these factors are not always obvious within the chart or graph. Figure 16 shows an example of what a leading and lagging metric relationship can look like. Tracking asset Downtime is beneficial, however, having a suite of metrics that precede downtime can help pinpoint the reason for success or failure within a maintenance strategy.

Harmonized Metrics Interactions	Maintenance $ % RAV	Maintenance Unit Cost	MRO Value % RAV	Maintenance Materials	Inventory Turns	Corrective Maintenance	CBM Cost	PM Cost	Maintenance Shutdown	Availability	MTBF	MTTR	Schedule Compliance	Stock Outs	Corrective Work	Reactive Work	Proactive Work	CBM Work	PM Work	Overtime	Result or Behavior?	Target Direction
Maintenance $ % RAV		Down	Down	Down	Up	??	??	??	Down	Up	Up	Down	Up	Down	50%	Down	80%	15%	15%	12%	Result	Down
Maintenance Unit Cost	Down		Down	Down	Up	??	??	??	Down	Up	Up	Down	Up	Down	50%	Down	80%	15%	15%	12%	Result	Down
MRO Value % RAV				:	Up	:	:	:	:	:	:	:	:	:	:	:	:	:	:	:	Result	Down
Maintenance Materials Cost					Up	Down	??	??	Down	Up	Up	:	:	:	50%	Down	80%	15%	15%	:	Result	Down
Inventory Turns						:	:	:	:	:	:	:	:	:	:	:	:	:	:	:	Result	Up
Corrective Maintenance Cost							:	:	Down	Up	Up	Down	:	:	50%	Down	80%	15%	15%	12%	Result	Down
CBM Cost								:	??	:	:	:	:	:	:	:	??	??	??	:	Result	Up
PM Cost									:	Up	Up	:	:	:	:	:	??	:	??	:	Result	??
Maintenance Shutdown Cost										:	Up	Down	Up	Down	50%	Down	80%	15%	15%	12%	Result	Down
Availability											Up	Down	Up	Down	50%	Down	80%	15%	15%	:	Result	Up
MTBF												Down	Up	Down	50%	Down	80%	15%	15%	:	Result	Up
MTTR													:	Down	50%	:	:	15%	:	:	Result	Down
Schedule Compliance														:	50%	Down	80%	15%	15%	:	Behavior	Up
Stock Outs															:	:	:	:	:	:	Result	Down
Corrective Work																Down	80%	15%	15%	:	Result	Down
Reactive Work																	80%	15%	15%	:	Result	Down
Proactive Work																		15%	15%	:	Behavior	Up
CBM Work																			15%	:	Behavior	Up
PM Work																				:	Behavior	Down
Overtime																					Result	Down

Figure 16 - Leading vs. Lagging Indicators

TRACKING COST AVOIDED

Sometimes it is important to show the value of Preventive Maintenance in terms of cost, but this is easier said than done in most organizations. The information for Downtime is in one system, maintenance labor and materials Cost are in another. A simple calculation can show the avoided impact of preventive maintenance efforts if the preventive tasks are failure based, as discussed previously.

If all preventive maintenance tasks are failure based, not completing them <u>will</u> result in unplanned loss of function, or unplanned downtime. However, if the failure is detected prior to the point of functional failure, the "F" on the P to F curve, then a repair work order can be written and executed without impacting the operational requirements of the asset. The amount of time it took to complete that repair, for example 3 hours, is, at a minimum, the amount of unplanned downtime that was avoided. For this reason, it is possible to calculate the amount of cost saved in downtime hours for each PM, or the entire program. Taking it to the next level, if the amount of profit per hour is known, that dollar amount can be applied to the result as well.

$$Cost\ Avoided = (Total\ Repair\ Labor\ Hours) * (Profit\ per\ Hour)$$

$$\$320,000 = (320\ Hours) * (\$1,000\ per\ Hour)$$

Remember, it is important to train users on the importance of data in the CMMS. Failure data and work order specific data, such as actual labor hours and actual parts and materials, are so important when evaluating the impact of Preventive Maintenance Optimization. Without this data, it is nearly impossible to show the journey while you're waiting for the results. The results never tell the entire story. It's all about the journey.